ULTIMATE SUPERCARS

PORSCHE 911 GT2

By John Perritano

Kaleidoscope
Minneapolis, MN

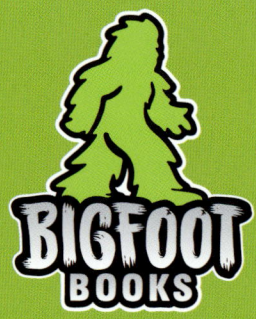

The Quest for Discovery Never Ends

....

This edition first published in 2021 by Kaleidoscope Publishing, Inc.

No part of this publication may be reproduced in whole or in part without written permission of the publisher.

For information regarding permission, write to
Kaleidoscope Publishing, Inc.
6012 Blue Circle Drive
Minnetonka, MN 55343

Library of Congress Control Number
2020936109

ISBN
978-1-64519-267-1 (library bound)
978-1-64519-335-7 (ebook)

Text copyright © 2021 by Kaleidoscope Publishing, Inc. All-Star Sports, Bigfoot Books, and associated logos are trademarks and/or registered trademarks of Kaleidoscope Publishing, Inc.

Printed in the United States of America.

Bigfoot lurks within one of the images in this book. It's up to you to find him!

TABLE OF CONTENTS

Chapter 1: Road America 4

Chapter 2: A Luxury Icon is Born 12

Chapter 3: A Miracle of Engineering 16

Chapter 4: Grandpa's Ride 24

Beyond the Book 28
Research Ninja 29
Further Resources 30
Glossary 31
Index 32
Photo Credits 32
About the Author 32

Chapter 1
Road America

The video looks like a game. It doesn't look real. But it is. If you like speed, you have to see this!

David Donohue is behind the wheel of a red-and-black Porsche 911 GT2 RS. It's April 2019. Donohue is at Road America in Elkhorn, Wisconsin. Race fans and car lovers know all about the track.

FUN FACT
Porsche 911 GT2 cars take part in racing series in America, Europe, and Asia.

It's a challenging 14-turn course. It's one of the best places to be to see how fast a car can move.

Donohue wants to go fast. That's his job. He's a race car driver.

A dash cam is mounted inside the GT2. You can see a graphic of the car's dashboard. Its throttle. Its brake. Its **tachometer**. Its gears and speedometer. You watch as Donohue throws the car in gear. Off the car goes. There's a map of the track so you know where the car is at all times.

PARTS OF A
PORSCHE 911 GT2

You stare at the speedometer. You're watching the numbers move.

162 miles per hour.

The GT2 is still moving. No, that's not right. The car is flying.

170.

You wonder how fast this car can go.

172.

Then the GT2 RS hits 179. That's 179 miles per hour.

Air intake scoops

The GT2 roars around a track near a forest. Racers around the world know the power of the Porsche!

Rear spoiler

Cooling water spray

Different sized front and rear tires

You're amazed! You're puzzled! Did you just see what you just saw? And that's not even close to the car's top speed of 211 mph (340 kph).

Donohue enters a curve. He **downshifts**. The engine scales back. The car slows.

IN THE DRIVER'S SEAT

David Donohue is a racer like his dad, Mark. David raced many types of sports cars in his career. In 2009, the younger Donohue won the Rolex 24 at the Daytona International Speedway. Donohue drove one of the Porsches that day.

Another straightaway emerges. Your chest thumps again.

This time the speedometer stretches back to 176.

More curves. More downshifting. The tach relaxes and you're finally able to take a breath. It was a wild ride and you weren't even sitting in the car. Donohue just broke a new **production car** record. It took GT2 RS two minutes and 15.17 seconds to speed around the four-mile track.

Donohue wasn't shocked. Earlier he had set a record at Road Atlanta in Georgia. The GT2 ripped around the track in 1 minute 24.88 seconds. It beat the 911 GT3 RS by 1.36 seconds.

Speed! It's why a person buys a Porsche. No one does speed better than the German company. Ferrari is very fast. Maserati makes fast cars, too. There are lots of other speedsters.

Yet there's something about a Porsche 911 GT2 RS. It has crazy power. It's packed with style. It has class.

It's a Porsche, after all.

This is the regular GT2, made for street driving.

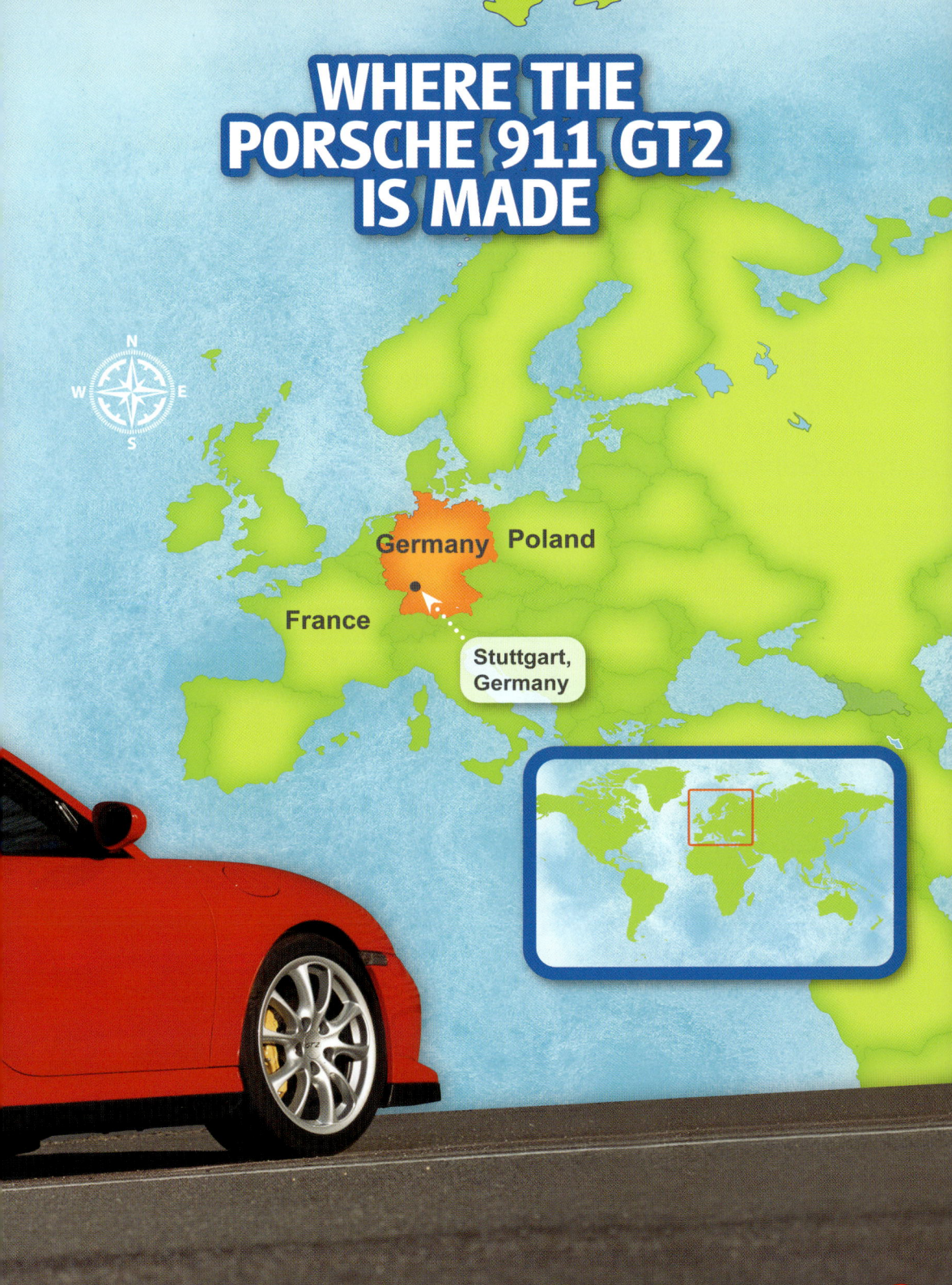

Chapter 2
A Luxury Icon is Born

Porsche sits on top of the sports car pyramid. Today's Porsche company began in Stuttgart, Germany in 1948. That firm had 200 workers. Its history dates back even further.

Ferdinand Porsche was born in 1875. He worked with his father as a plumber. The German government in the 1930s asked Porsche to design a car for the masses. Porsche came up with the Volkswagen Beetle. Volkswagen means "people's car."

World War II began in 1939. Porsche designed tanks for the German army. He opened up his own car factory after the war.

The company's first car was the Porsche 356. Along came the 550 Spyder in 1952. It was lightweight and sleek. Drivers loved it.

Then in 1964, Ferdinand made the 911. It became the face of Porsche.

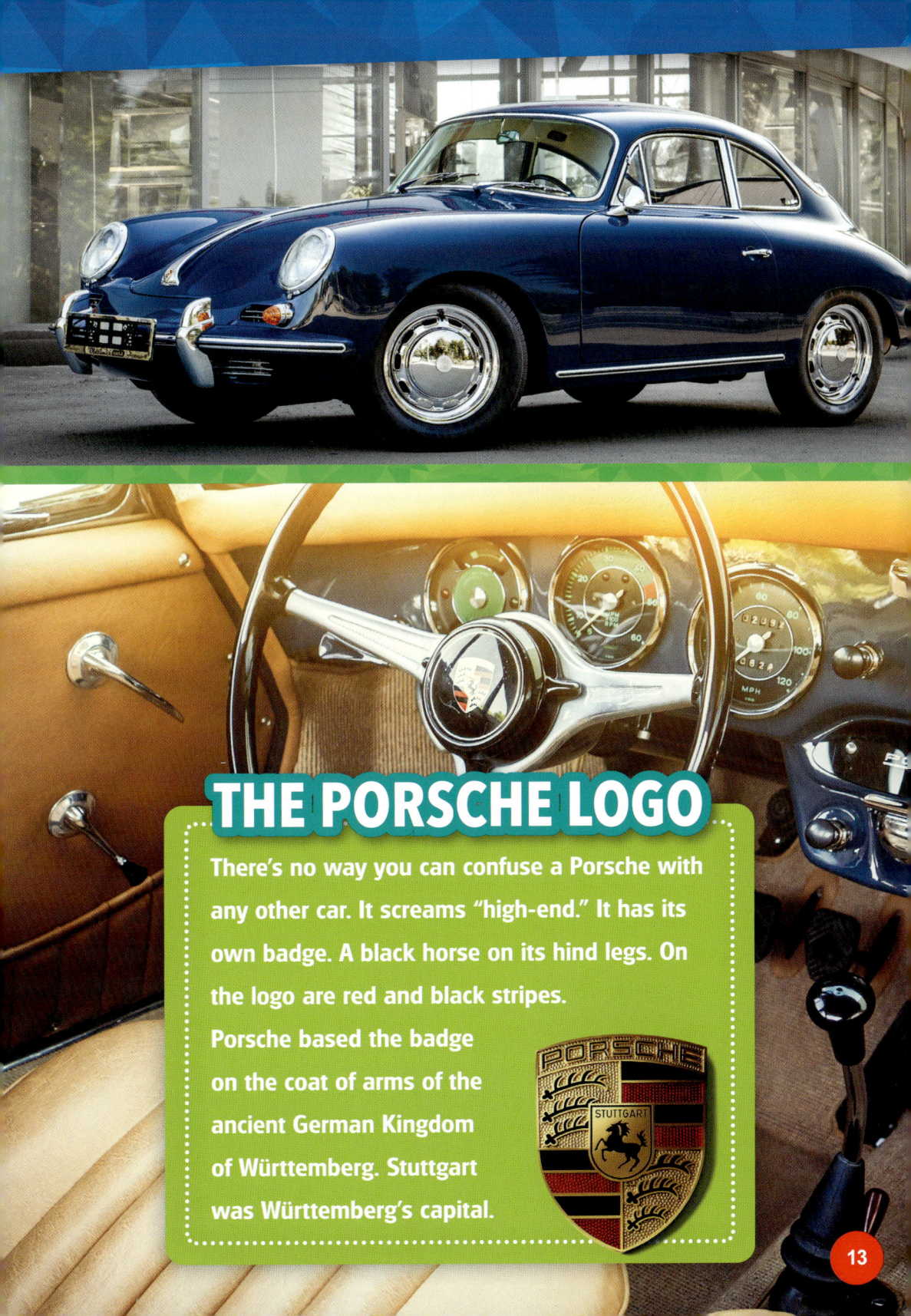

THE PORSCHE LOGO

There's no way you can confuse a Porsche with any other car. It screams "high-end." It has its own badge. A black horse on its hind legs. On the logo are red and black stripes. Porsche based the badge on the coat of arms of the ancient German Kingdom of Württemberg. Stuttgart was Württemberg's capital.

The Porsche 911 has always been one of the world's fastest cars.

In 2008, Porsche introduced its 911 Turbo. It was the first gasoline-powered production car to use an engine design called VTG. This design made the engine more powerful than ever.

The 2008 911 had the classic Porsche form. It was popular on the street and the track.

The 911 had other major changes that increased its performance. It had a longer wheelbase (the width of the car). A new **suspension system**. A seven-gear manual **transmission**. Larger tires.

The 911 Turbo was terrific. However, the GT2 took the 911 brand to new heights—and new speeds.

Chapter 3
A Miracle of Engineering

Amy's grandpa Jack always liked fast cars. That's how Jack rolled. He loved to tell stories about his cars. He always told about how he drove a 1967 Mustang Shelby from Chicago to San Diego.

"Sometimes we'd open her up just to see what she could really do . . . she did a lot," Jack said.

Grandpa Jack was only a teenager then. As he grew older, his love for fast cars never faded. He had a few over the years. What was his favorite? Jack pointed. "That one," he said, pointing to the GT2 RS. "I always loved the Porsche GT2."

The world got its first glimpse of what would become the GT2 at a Geneva Motor Show in Switzerland.

At Geneva, Porsche put the 911 Turbo on display. It was on a 964 **chassis**. It had rear-wheel drive and was monster powerful. It wasn't heavy. That meant less traction. It was a recipe for a great car.

Porsche officially introduced the GT2 in 1995. It had a 430-horsepower engine. Engineers later pumped the power plant up to 450 HP. They took the car to a legendary track in Nordschleife, Germany. There the car set a lap record of 7 minutes, 56 seconds.

KEEPING ON TRACK

The Nordschleife track in Nürburg, Germany, is a speedway like no other. Built in the 1920s, it's a special place for race drivers. The track is very demanding. There are steep hills. Drivers have to make dangerous corners. The road surfaces are tricky. Famous cars and drivers both love and dread this track!

FUN FACT
Sports car tracks feature many twists and turns like this one.

THE
PORSCHE 911 GT2 RS
IN DETAIL

Height: 4 feet, 3.1 inches (1.29 m)

Width: 6 feet, 2 inches (1.8 m)

LENGTH: 15 feet, 6 inches (4.8 m)

WEIGHT: 3,241 pounds (1,470 kg)

TOP SPEED: 211 miles per hour (339 kph)

TIME FROM 0-60 MPH: 2.7 seconds

COST: $293,000 (United States)

In 2002, the GT2 came to America as the 996. It even looked fast! There was no rear seat. That reduced the car's weight by 17.6 pounds. Porsche took away the sunroof. It replaced the full spare tire with a reinflation kit. That saved 29 pounds. Engineers replaced the steel braking system with carbon ceramic. That saved more weight.

Now lighter in weight, the 996 GT2 handled better. The back had a wider rear track. That added more stability. It had bigger wheels and bigger tires.

Engineers put a large rear wing on the back. The wing created more downforce. Downforce pushes down on a car. It allows the tires to grip the road better.

All the controls in the GT2 are designed for high speed.

By 2004, Porsche bumped up the GT2 to 477 horses. A new computer system delivered the fuel better. But Porsche wasn't done yet. In 2007, they introduced a GT2 with 530 horses. It powered around Nordschleife in Germany at 7 minutes, 32 seconds.

Porsche added RS to the GT2's name in 2010. That stands for "racing sport." With the new name came more horsepower. The GT2 RS now boasted 620 "ponies." (That's a nickname for horsepower, get it?) The new model's suspension system was more rigid. That helped drivers control the higher speed and power.

WHAT, NO RADIO?

Speed is the name of the game for Porsche. That's why the GT2 is slim and trim. Less weight equals more speed. Porsche will even sell its GT2 without air conditioning. The company will even remove the radio. Every pound counts. You probably wouldn't hear the music anyway. The GT2 RS is too loud.

Chapter 4
Grandpa's Ride

"Get in," Grandpa Jack says to Amy.

The interior of Jack's GT2 is just as cool as the exterior. He pulls out of the driveway. He takes the car on a lonely stretch of road.

"Hang on," he smiles.

FUN FACT
The spoiler at the back of the GT RS helps keep the car stable at high speeds.

Grandpa Jack presses the gas pedal. His GT2 RS takes off. Gravity pushes them both back in their seats. The car hits 60 mph in 2.7 seconds.

"If we were on a track, she'll do 211," Grandpa says. Amy smiles. She's never gone that fast!

The GTS is aerodynamic. That means it cuts through the air very cleanly. Its side skirts are wider. That adds to the downforce.

"I don't think you would find a more powerful engine," Grandpa says. "Just listen to it roar."

The 911 GT2 RS in 2017 became the fastest 911 of them all. That's what 700 horses will do for you.

Porsche found more ways to save weight. The car's roof is made of magnesium. The exhaust system is titanium. The 911 GT2 weighs only 3,241 pounds (1,470 kg).

The power is incredible. Jack takes it on the highway. The guardrails zip by.

"Lars Kern drove this same model at Nürburgring," Grandpa says. "One lap took him only 6 minutes and 47.3 seconds." That was a new record for a non-racing car.

Jack wraps up the ride. He glides the car back into his garage. Even Porsches need to take a rest sometimes!

Ready to race!

The six red tubes show the six cylinders of the 911 GT2. They put out more than 600 horsepower to create speed!

BEYOND
THE BOOK

After reading the book, it's time to think about what you learned. Try the following exercises to jumpstart your ideas.

RESEARCH

FIND OUT MORE. Where would you go to find out more about your favorite cars? Find out what company makes the car and locate its website. What information do the companies provide? What other sources of car information can you find?

CREATE

GET ARTISTIC. Cars start with creative artists and designers. Time for you to take a shot! Get art materials and create a great, new car. Will you make it a sports car? A sedan? A race car? What colors will you paint it? What features can you give it? Let your imagination go for a spin!

DISCOVER

DIG DEEPER. Ferdinand Porsche built amazing cars. He also worked for his home country of Germany during World War II. Find out more about his military work. Does it change how you feel about him? What about how you feel about Porsche cars?

GROW

GO TO A CAR SHOW. Car shows are a great way to see lots of cool cars up-close. Check your local events calendar, or ask at a car dealer for upcoming events. You can find shows of old cars and new cars, sports cars and classic cars. Go to a show and find a new favorite car to love!

RESEARCH NINJA

Visit **www.ninjaresearcher.com/2671** to learn how to take your research skills and book report writing to the next level!

RESEARCH

DIGITAL LITERACY TOOLS

SEARCH LIKE A PRO
Learn about how to use search engines to find useful websites.

FACT OR FAKE?
Discover how you can tell a trusted website from an untrustworthy resource.

TEXT DETECTIVE
Explore how to zero in on the information you need most.

SHOW YOUR WORK
Research responsibly—learn how to cite sources.

WRITE

GET TO THE POINT
Learn how to express your main ideas.

PLAN OF ATTACK
Learn prewriting exercises and create an outline.

DOWNLOADABLE REPORT FORMS

Further Resources

BOOKS

Gastecki, Julie. *Porsche 911.* Mankato, MN: Black Rabbit Books, 2020.

Leffingwell, Randy. *Porsche: 70 Years.* Minneapolis, MN: Motorbooks, 2017.

Leffingwell, Randy. *The Complete Book of Porsche 911.* Minneapolis, MN: Motorbooks, 2019.

WEBSITES

FACTSURFER

Factsurfer.com gives you a safe, fun way to find more information.

1. Go to www.factsurfer.com.
2. Enter "Porsche 911 GT2" into the search box and click 🔍
3. Select your book cover to see a list of related websites.

Glossary

chassis: frame of a motor vehicle.

downshift: a change into a lower gear in a motor vehicle.

production car: motor vehicles that are mass produced for sale to the public.

suspension system: the system of springs and shock absorbers connecting the wheels to the chassis of a motor vehicle.

tachometer: instrument that measures the speed of an engine usually in revolutions per minute.

transmission: the mechanism through which power in an automobile is transmitted from the engine to the wheels.

Index

550 Spyder, 12
Daytona International Speedway, 8
Donohue, David, 4, 8, 9, 10
Donohue, Mark, 8
engine, 8, 14, 18, 25, 27
Geneva Motor Show, 17, 18
Georgia, 10
Germany, 11, 12, 13, 18, 22
Kern, Lars, 26
Maserati, 10
Mustang Shelby, 16
Nordschleife, 18, 22, 26
Nürburgring, 26
Porsche, Ferdinand, 12
Road America, 4
Road Atlanta, 10
Rolex 24, 8
spoiler, 7, 24
Volkswagen, 12
Wisconsin, 4
World War II, 12
Württemberg, 13

PHOTO CREDITS

The images in this book are reproduced through the courtesy of: Courtesy Porsche Media Center: 4, 8, 10, 14, 19, 21, 22, 24, 26. Shutterstock: Lawrence Carmichael 6; Fariz Abasov 20; The Highest Quality Images 7; Everyonephoto Studio 12; Gaschwald 14 inset; VanderWolf Images 16. **Cover:** jvdwolf/123rf (car); Svetlana Foote/Shutterstock (background, top); zhao jiankang/Shutterstock (background, bottom).

About the Author

John Perritano is an award-winning journalist, author, and editor from Southbury, Connecticut. He has authored numerous books and articles on subjects such as science, technology, history, and current events. He holds a master's degree in American History from Western Connecticut State University.